To [illegible],
my friend
Enjoy!
Love,
Frank

THE SUN (SON) RISES ABOVE THE EARTH'S AURA

FRANK YOUNG

The New American Bible, Catholic Bible Publishers, Wichita, Kansas, 1975-76 Edition

The Holy Bible, Good Counsel Publishers, Chicago, Illinois, J.G. Ferguson Publishing Company 1960, 1963, and 1965

Holy Bible, authorized King James VERSION, The World Publishing Company, Cleveland and New York 2231 West 110th Street, Cleveland, Ohio.

LifeRich Publishing books may be ordered through booksellers or by contacting:

LifeRich Publishing
1663 Liberty Drive
Bloomington, IN 47403
www.liferichpublishing.com
1 (888) 238-8637

ISBN: 978-1-4897-1140-3 (sc)
ISBN: 978-1-4897-1139-7 (e)

Print information available on the last page.

LifeRich Publishing rev. date: 01/26/2017

DEDICATION OF BOOK TO MY SPOUSE ROSEMARY

"I shall always think of you wandering through a lovely garden, like that which you fashioned with your own hands, where flowers never fade and no cold wind of sorrow blights our hopes and plans – and on your face the peace of one whose whole life through, walked with God." (Bessie Morse Bellingrath 1878 – 1943)

This is how I think of my lovely and blessed wife, Rosemary, who I describe as "The Spice of My Life". She is a cherub who spreads good will and charity in her daily actions and love and caring with everyone she interacts with. She never meets a stranger.

In the beginning God created the heavens and the earth

This preamble in the Bible is paramount to understanding the relationship between God and nature. It is a pronouncement which clearly states there is a Creator, a natural world on which we dwell, and an intrinsic association between The Creator and that worldly habitat. It is the most basic response to that most fundamental question – how? How did Earth and Nature come to be? How did the globe with all it entails, with all its splendor, evolve? The simple and obvious answer should logically be that it was constructed by God's hand and mind.

Contents

Introduction

My purpose in writing this concise book is to encourage the reader to think, to think about what should be the more important things in life, about something besides materialism such as houses, cars, clothes, lavish vacations, cushy jobs, status, prestige, pride, and upstaging the Jones'. My goal is to convince you, The Reader, to take the time to "smell the roses", witness the miracles of nature, and permit science to be your best evidence of a higher being without interfering with your current or preconceived notions about your Creator. Your beliefs and your image of God should be left in tact, your faith reinforced by all that you see, hear, feel, and experience in a world made for you by your Creator, not likely the result of some *big bang* or evolutionary theory of biological evolvement, but the product, the result if you will, of a mastermind who meticulously invented, designed, and cultivated the Blessing we all share which is Life.

So, without my burdening you with volumes of verse and verbiage, let me stimulate your thought process in as few words as possible and share with you common questions, experiences, and observations. Just as the Bible rambles on without apparent literary transition, I will follow the writing style of the prophets who wrote each scriptural proverb and verse in the Greatest Story Ever Told [The Bible]. I'll be as brief and succinct as possible in my presentation by offering up phrases, quotations, excerpts from the Bible as well as from other authors or writers, random and provoking thoughts, speak in parables as did Jesus when addressing the Assembly, and embellish the text using little periodic snippets which may stimulate your thought process. I will not engage in an esoteric, deeply profound, essay which inundates you, the Reader, with overly verbose language or redundancy. There will be no chapters, glossary, traditional footnotes, references, or other standard book material.

Hopefully, we'll reach the consensus that there is something bigger than us and our perceived reality. Perhaps we'll recognize that we are all here for a reason and that we share a world that we can identify with, and somewhat comprehend, in the context of the mystery of what we understand or don't understand. In this common adventure we can be comfortable in the knowledge that there is a logical sequence, a purpose, and a rationale to our being who we are and where we are at this time. As I ask open-ended questions, moving forward, we will all seek answers, not knowing anything for sure, but moving in a positive direction that allows us to focus on the gift that God gave us devoid of the distractions of survival and everyday strife.

So while I don't presume to know the answers to my questions, this experiment will be enlightening to all of us in freeing up our minds, hearts and souls to the love, kindness and purity of the Holy Spirit and allow all the aforementioned to enter and dwell within us forever and ever.

Speaking of the Bible, the word Covenant appears 289 times in the New American Bible from Genesis to Revelation. The idea of covenant is the golden thread weaving in and out through the Bible's many books. The Church's approach to the Bible has long encouraged us to seek a broader understanding of biblical writings, what Scripture scholars call the sense of Scripture. When we look at Covenant that way, it is evident that God has always desired humanity to enter freely into an ongoing relationship of love with Him. As Believers we do not take the pre-historical chapters of Genesis (1-11) as literal history. Prehistory is that period which existed before written records were made and where the authentication of certain episodes (creation, the flood, the Tower of Babel) is difficult to establish. However, we do see in these accounts great truths which have had profound influence on every generation throughout salvation history. The last book, the Book of Revelation, while bewildering, visions the Golden Thread coming to rest at last at the Throne of God.

> *For the invisible things of Him, from the creation of the world, are clearly seen, being understood by the things that are made; His eternal power also, and divinity.*
>
> **–Romans 1:20**

The Dogwood - the Cross of Our Lord

According to Christian tradition, dogwood blossoms bear the scars of Jesus' crucifixion. The petals are tipped with blood-fringed nail indentations, the flower is cross-shaped, and the center represents the crown of thorns. This symbolism is to compensate for the dogwood being the wood of the cross on which Jesus died.

Because the tree seemingly never wanted to be put to such use again, Jesus spoke to it saying it would no longer grow large enough to be used for cruel purposes ever again and therefore God twisted and gnarled its trunk and branches so that no straight boards could be made from it. While there is no evidence that this is true, the story persists. It is important to stress that the legend of the dogwood blossom is an example of a tale that attempts to explain why something has its particular features or characteristics with no basis in scientific fact.

However, in most cases, the dogwood tree is used to signify the crucifixion of Jesus Christ. And even though the Bible does not talk about the type of tree on which Jesus was crucified, traditional stories or legends claim he was nailed on a dogwood tree.

Furthermore, the plant supposedly used to grow into a very large tree and because of this, it was mainly used to make crosses.

The Passion Flower – Christ's Thorns

This colorful flower is replete with religious reverence. Called "Christ's Thorns" by Spanish Christian Missionaries, who first discovered it in South America, each part of the flower embraces symbolic meaning in recognition of the crucifixion story – the Passion of Christ. Namely, five sepals and five petals refer to the ten faithful apostles (excluding Judas and Peter). Three stigma represent the three nails that held Christ to the cross; while five anthers represent his five sacred wounds. The tendrils of the flower are said to resemble the whips used in the flagellation, while the filaments, which can number in excess of a hundred, depict the crown of thorns. This powerful symbolism has led to the inclusion of the Passion Flower among the ornamentation of various churches, such as in stained glass window designs, altar frontals and lectern falls. It is also contended that The Passion Flower is sacred outside the Christian world.

In India the flower's structure is interpreted according to the story of the five Pandava brothers with the Divine Krishna at the center, opposed by the army of one hundred at the outside edges. The pigment of the blue Passion Flower is argued to be associated with the color of Krishna's aura.

Caterpillar to Butterfly

The Metamorphosis of the Caterpillar Into the Butterfly

> "Just when the Caterpillar thought the world was over, she became a Butterfly.
>
> The caterpillar dies so the butterfly could be born. And, yet, the caterpillar lives in the butterfly and they are but one."*

The above metamorphosis tells us that we live an earthly life and then ascend into Heaven.

*(**Proverbs Way Blog)**

Dove

Doves, which are in fact domesticated rock pigeons, are a well known Christian symbol of love and peace.

The story of Noah's Ark is one of the best examples of how the dove symbol was integrated into Christian history as a symbol of peace.

In the biblical story, a dove was released by Noah after the flood in order to find dry land. The dove returned carrying an olive leaf in its mouth, which confirmed to Noah that there was, indeed, dry land nearby.

Ever since then Christians have used the dove as a symbol for peace. There have been other stories that predate Noah's ark and equate the dove with peace, but the most important beginning of the dove as a religious symbol was, to reiterate, Noah's ark story.

In Christian beliefs, a dove also symbolizes the Holy Spirit, this substantiated by a couple of important biblical stories where the Holy Spirit is shown as a dove descending from heaven at the baptism of Jesus.

Furthermore, in Rome, early Christians incorporated the dove into their funerary art, the image of a dove carrying an olive branch often joined by the word "Peace" was used.

In fact, it would seem that when early Christians derived this symbolism from the Gospels, they combined the dove with the olive branch thereby reflecting peace by both Greeks and Romans.

Bear in mind too, that the Dove is a symbol of the Catholic sacrament of Confirmation.

Finally, doves are sometimes released at Christian weddings as a symbol of taking flight. It is a beautiful gesture which symbolizes a new beginning of something wonderful.

> ""After being baptized, Jesus came up immediately from the water; and behold, the heavens were opened, and he saw the Spirit of God descending as a dove {and} lighting on Him" [Mt 3:16]

> Matthew 3:16 Matthew 3 Matthew 3:15-17 "As soon as Jesus was baptized, he went up out of the water. At that moment heaven was opened, and he saw the Spirit of God descending like a dove and lighting on him."

Christian Symbolism: The Natural World

A list of the elements of nature that appear in the Bible, compiled from the New International Version (NIV) of the Holy Bible, contains 542 words. Around 70% of the words are universally-recognized, everyday things like land, water, fire, earth, and the natural division of time, a day, a month, or a year. The most frequently used words are day, land, year, earth, water, gold, sea, silver and night.

When all 542 elements of nature are grouped into general categories, the largest groups are animals (24%), plants (16%), geography (16%), and terms referring to water (13%). Relatively smaller groups are land (11%), environment e.g. hot, dark, blue, morning (10%), minerals (5%), earthly heavens, e.g. sky, space (5%), and terms referring to fire (1%).

The order in which the Bible tells us that the elements of nature were created by God is:

First day

Heaven, earth, water, light, and darkness with their names of day, evening, and morning

Second Day

Sky

Third Day

Land, seas, and vegetation

Fourth Day

Seasons, days, years, sun, moon, and stars

Fifth Day
Sea creatures and birds

Sixth Day
Land creatures and livestock

References of the following in the Bible and in Scripture:
Drought
Rain
Flood
Snow
Lightning
Thunder
Storm
Hot and cold
Water, Light, Dove, Wind
Birds, Sky, Clouds
Animals
Earthquake
Hail, Fawn, Horse, Camel, Lion, Sheep, Goat, Tiger, Bear, Cow, Elephant, Deer

Nature can be grouped into 8 categories:

- Light
- Water in its many forms like fog, ice, rain, rivers
- Sky as everything above the land including air, wind, space, stars
- Land
- Plants
- Animals and things that are unique to animals like a wing, roar, hoof
- Fire including things that fire produces, like ash and smoke
- Decay

> Genesis 1:24 Genesis 1 Genesis 1:23-25 "And God said, 'Let the land produce living creatures according to their kinds: livestock, creatures that move along the ground, and wild animals, each according to its kind' And it was so."

Man is like a bird. Got to keep on flowing where the forces of nature take you.

The Miracle of the yellow daisy, its seed become yellow flowers, or life, those seeds parachute and ascend to land, grow and perpetuate life. The old saying about popping up daisies is literally true in that where the dead lie, they grow as new life emerging.

"Pushing up daisies: death to life."

Plants

The plants found in sacred objects and images, for example the vines and palm trees, are often mentioned in the Bible. In other cases, they are linked to a particular event that is important in Christian belief.

Wheat

Ears of wheat symbolize the body of Christ and the bread taken during the service of Holy Communion. During the Last Supper, the last meal that Christ shared with his followers before his death, he gave bread to his companions and told them to eat it in remembrance of him. In the Roman Catholic tradition worshippers accept the bread as the body of Christ, while for many Protestants, Holy Communion is a symbolic commemoration of the Last Supper. Wheat is also a sign of God's bounty and of the harvest. It represents the word of God that grows on fertile soil.

Grapes and Vines

Grapes symbolize the blood of Christ and the wine taken during the service of Holy Communion. During the Last Supper, the last meal that Christ shared with his followers before his death, he gave wine to his companions and told them to drink it in remembrance of him. In the Roman Catholic tradition worshippers accept the wine as the blood of Christ, while for many Protestants, Holy Communion is a symbolic commemoration of the Last Supper. A vine and its branches also represent Christ and his followers.

Trees

> Job 14:7 Job 14 Job 14:6-8 "Fore there is hope of a tree, if it be cut down, that it will sprout again, and that the tender branch thereof will not cease."

> Psalm 1:3 Psalm 1 Psalm 1:2-3 "And he shall be like a tree planted by the rivers of water, that bringeth forth his fruit in his season, his leaf also shall not wither, and whatsoever he doeth shall prosper."

Why is it in nature that everything exists for a purpose? Why is it that everything happens for a purpose like a clock according to a time table? Why is it that leaves fall off trees when winter approaches? Because, if the leaves did not fall, the tree itself could not live throughout the winter months, for without the nourishment from the fallen leaves the tree would wilt and die. Why is it that trees (let's take a pine tree for example) have two types of pine cones? Their purpose for existing is to produce more trees until from one tree there emerges a forest. Why is it that a tree has ducts, the xylem and phloem? Because the ducts transport food and water throughout the tree. Why is it that trees have roots? Because to absorb moisture and soil nutrients from the earth, a tree must have roots. Why is it that a tree has bark? To protect the organism from parasites and inclement weather. And finally, why does the tree itself exist? Trees provide shade for the thoughtful, comfort for the weary, inspiration for the human mind, and lumber to house the human body. A tree is not just a happening, no evolutionary theory can explain its origin, no scientific theory for that matter can explain its origin. The tree, as anything else in nature or life is here for a purpose, a reason. A tree is a phenomenon like the earth and the stars, carefully constructed, carefully planned, on earth for a reason, on earth for a purpose.

Palm Branch

The palm is an ancient symbol of victory in war. It then became a symbol of Christ's victory over death. When Christ entered Jerusalem shortly before his Crucifixion, he came in triumph and his followers greeted him with palm branches. Later he was arrested and killed, but three days later he rose from the dead. Christ's Resurrection is the central belief of the Christian faith.

Animals

Animals are often mentioned in the Bible. In the book of Genesis, for example, the serpent tempts Eve to eat forbidden fruit in the Garden of Eden. For this, he is cursed and becomes a vile creature fated to crawl in the dust. Christian teaching uses the nature of animals to explain spiritual qualities. For example, in the face of death, Jesus surrendered to God's will with the meekness of a lamb.

Sheep and Lamb

Sheep represent vulnerable human beings who need the guidance, care and protection of Christ, the Good Shepherd. The 'Lamb of God' symbolizes the sacrifice that Christ made for humankind, since it was a Jewish custom to sacrifice lambs. The lamb also reflects Christ's innocence and the white fleece his purity. The lamb is sometimes shown holding a cross or a banner to signify Christ's victory over death through his Resurrection.

Birds

Birds and other winged creatures have come to represent certain aspects of Christian teaching. For example, the dove signifies the Holy Spirit and a dragon is a symbol of the devil.

Pelican
In legend, the pelican pecks at its breast to feed its young with its own blood if food is scarce. It is a symbol of unselfish love, charity and sacrifice, and therefore of Christ who gave his own life to save humankind from sin and death.

Sun and Moon

Genesis, the first book of the Bible, tells us that God made the sun and moon on the fourth day of creation. They are symbols of the power of light to shine through the darkness, as good will shine over evil.

The sun can also represent the radiance and glory of Christ as seen on the crucifix. In scenes of the Crucifixion, the moon is sometimes shown on the left of the cross and the sun on the right. This could refer to an early church teaching in which the Old Testament (moon) is lit by the New Testament (sun).*

***Victoria and Albert Museum**, Cromwell Road, London, SW7 2RL

Ichtus (Fish Symbol)

The fish as a symbol in Christianity is nearly as old as the Christian faith itself. The sign is seen in the past on things like art and architecture and today it endures on things like bumper stickers and business cards as a sign of Christian faith. The fish is thought to have been chosen by the early Christians for several reasons:

- The Greek word for fish (ICHTUS), works as an acrostic for I = Jesus, C = Christ, TH = God's, U = Son, S = Savior
- Jesus' ministry is associated with fish: he chose several fishermen to be his disciples and declared he would make them "fishers of men." (Also see the New Testament and the Book of Matthew)

The fish is also a symbol of baptism, since a fish is at home in the water.

"As Jesus walked beside the Sea of Galilee, he saw Simon and his brother Andrew casting a net into the lake, for they were fishermen. 'Come, follow me,' Jesus said, 'and I will make you fishers of men.' At once they left their nets and followed him." ~ Mark 1:16-18

Reality

- Reality is only what we can feel, hear, touch, smell, or see. Life is like a movie projector, e.g. cameras, light, action.

- What was non-life like before we were born? Can we ever remember? Can we expect to remember an existence devoid of life and light?

- How do we know that we are not peering into a Christmas Carol window, as per the movie Scrooge, seeing a vision, having a dream, part of a script?

- What is reality? When we're asleep and dreaming or awake and walking around?

- "Relish the moment" is a good motto, especially when coupled with Psalm 118:24 "This is the day which the Lord hath made; we will rejoice and be glad in it." To reiterate, it isn't the burdens of today that drive people mad. It is the regrets over yesterday and the fear of tomorrow. Regret and apprehension are twin thieves who rob us of today.

- Close your eyes and then reopen them. Reality is only what we see and experience. Big houses, cars are only enjoyed while the dream is lived.

- Snap, crackle and pop in someone's cereal bowl. Are we being watched by higher life forms? We have to assume that God is watching all of us as we move about the globe, right?

- Life is like watching TV. It is like a series of film clips which we can freeze or keep reeling.

- If the future is hard to imagine, then the past, while hard to reconstruct or visualize, is equally likely to be real. Why must we see to believe? A photo is evidence of the past.

- We look at the past (or into it) and specifics or details can be remembered but they always remain vague even as we try to recall a dream only a few minutes old. Some dreams we can't recall at all. Some we don't want to recall.

- We can't see ourselves as others do in three dimension. We cannot see our backs, the backs of our heads, our own faces. We see ourselves in mirrors, but never total reality, never in the flesh, as we see others. We are denied seeing ourselves except, second hand, through pictures and film, luxuries people centuries ago were denied.

- Sleeping and awakening is based on faith that this will happen like clockwork. Darkness and daylight are predicated on faith.

- Death and sleep? Perhaps we are dead or in a sleep before we are born to life and light

- Could life be like a dream that we turn off when we sleep instead of vice versa?

- We freeze the dream when we make films, take pictures, or paint portraits.

- We freeze time in pictures like ice preserves prehistoric life.

- Man tries to make regular what is irregular. Man tries to make logical what is illogical from the standpoint of being beyond logic or incomprehensible, beyond his scope of understanding. Are we dust in the wind? Do we derive from dust if we become dust? Are we not like the flower, or plant, that originates from the soil, dies and regenerates from the earth? Are we like plants and conform, like them, to a natural cycle?

- Change is so rapid. Even as we speak, or even as you read these words, you are changing physically, chemically, and emotionally. People, if you watch them from high up, look like ants, so to speak, particularly if you view them from a tall building and watch them travel about the expressway system, watch them as pedestrians in shopping centers, college campuses, or on the streets. People are like ants or bees or like microscopic life, interchanging, merging, remerging, fusing, fission, separation, part of a systematic yet maddening whirlpool of inexhaustible activity. We look at ourselves as so important yet we are probably no more significant than the so-called lower forms of life we observe on the ground or via the microscope. We are probably being viewed by higher forms of life and certainly by God as we go about our frenzied way of life down the fast lane.

- Would a fish ever feel wet?

I Say to Myself "What a Wonderful World!"

Some of my support blocks:

- I contrast myself to others less fortunate, e.g. to the crippled, the deformed, the mentally ill, the indigent, the incarcerated, and am thankful that I'm not less fortunate.

- I look back on where I've been and how far I've had to come to be standing where I am now, no matter what the level of my success.

- I don't compare myself to the achievements of celebrities, the affluent, company executives, or others in that I have to live my life cognizant of my own limitations, and with the knowledge that I don't have aspirations to be something more, complacent with what I am at the moment. I don't let peer pressure impact me if I can help it. I'm a maverick.

- I try to always turn a negative experience into a perceived positive for moving forward and attempt to learn from life's mistakes.

- I thrive on hope as my harvest for the future.

- I remind myself that my priorities in life should be my spiritual life and how I treat others, i.e. whether I'm good in God's eyes. It sometimes takes life's bumps in the road that wake us up to the fact that debts and bills, material things like cars, houses, and frill money are of no intrinsic value. If we are sitting in a hospital room with no prospect for the present, or future, we realize then the superficiality of everything else. We don't want regrets, if we can help it, when we reach that crisis stage in life.

- I count my blessings which include my wife, my friends, my health, and the small things in life which mean more than riches. I'm thankful for my humble vocation and that I'm able to move around and discover the earth's wonders.

- I take time to smell the roses, the petunias, the carnations, the honey suckle, to feel the wind against my face, to hear the cricket, the dragon fly, the Jews harp, singing birds, the rustle of leaves on the trees, the racing of little animals in the forest. I take time to interpret the shapes of white billowy clouds against a blue backdrop, or to listen to the rumble of distant thunder with flashing lightning during a spring or summer rainstorm. I observe a bumble bee suck nectar out of a flower, watch a humming bird come to my feeding station, or see a butterfly gracefully dart around in its attempt to light somewhere. I'm like a little kid as I watch snowflakes gracefully fall from Heaven in winter, like manna from Heaven, each with its own fingerprint. I marvel at the rich crimson, gold, and orange of fall colored trees, the brilliance of dogwood and Bradford pears glistening like snow, and the beauty of azaleas in the spring as a robin searches for worms on the ground while geese fly over my head. These are God's gifts to us. My mother, a woman who I called Tiny (she was a little over five foot tall) would wake up each day and say, "This is a day to put in your pocket."

- Life is very precious and I'm awakened to this reality by movies like "It's A Wonderful Life" with Jimmy Stewart and Donna Reed, or "Far and Away" with Tom Cruise and Nicole Kidman.

This is why I cherish the song by the late Luis Armstrong called "I Say to Myself, What a Wonderful World".

- Don't worry about tomorrow when you can live today.

- Don't bemoan the past when you have hope for the future.

- Don't get ahead of yourself, but step one step at a time. Don't run through life, but walk and pause at each fork in the road.

- Sometimes it's harder to live than die it seems. Work at living. Don't quit.

- There are no guarantees in life.

- Always look at how things could be worse than better.

The Miracle and Mystery of Creation

The sun, moon, stars can be seen from anywhere, from any angle, from the four corners of the earth, or any place on the globe. This tells us that God's creations are there for all times, forever, and watching over us, just as God is omnipresent and watching over us.

Literally walking on water: Drs. Ray Noblet, Darold Batzer, Bob Matthews, Wayne Berisford, and Mark Brown, professors of entomology at the University of Georgia, informed me that there are two bugs that can tread on water; namely "There are two that come to mind immediately. One is a bug called the water strider; it has fairly long skinny legs and scoots around very rapidly. The second is a small round shiny black beetle called a whirligig. These are the two most common in this region." This would lead us to believe that Jesus did in fact walk on water, particularly if He were God's Son, if were God. He could do anything physically or spiritually.

The corporeal and the spiritual. We can lose body parts but the spirit of the heart and mind live on.

A seed in nature is like the womb in a woman. From it comes Life. Thus, "What a seed lacks in interest, it makes up for in miracles" is one of my favorite quotations.

> "Dreams are the seeds of change. Nothing ever grows without a seed, and nothing ever changes without a dream." *Author: Debby Boone*

> "Happiness held is the seed; Happiness shared is the flower." Sydney J. Harris author

The sounds and sights of nature are like the voice of God.

We must believe in Miracles for Life is a Miracle. It is normal for water to change into something resembling a rock when it freezes. We also know that, on its own, water never suddenly changes into wine, nor does a body of water open a path of dry land before us when we need to head across it. A bush does not burst into flame without incinerating. The sun and the moon never stop moving across the sky. A few small fish and 7 loaves of bread will not feed over 4,000 hungry people. When God makes such things happen, because of how familiar we are with what is normal in nature, we have no trouble recognizing that a miracle has happened. Over 70% of the miracles in the Old Testament and over half of all the miracles in the Bible are of this type – an altering of the normal ways of nature

The Cycle of Life

- We live life in four seasons, namely, birth and death in four seasons:

  Spring, new life

  Summer, zenith of life

  Fall, nostalgic/reflection, old age

  Winter, temporary death till spring

- We live and die each decade, a decade characterized by a certain aura, era, environment, diversity, demographics, e.g. 50's, 60's, 70's.

- Everything in life part of a cycle.

- Life is a wheel in motion, moving in circles.

- We live life in stages and then are reborn:

  Infancy

  Adolescence

  Middle-age

  Old-age

- From death can come Life
- Life is a series of phases, many lives rolled into one, we die and we are reborn.
- Pie in Geometry is 3.14, e.g. infinity; life is a circle, never ending.
- For everything there is a season.
- All things belong to him/her who waits.
- Salmon and penguins each reflect cycles of life.

Why is it in nature that everything exists for a purpose? Why is it that everything happens for a purpose like a clock according to a time table? Why is it that leaves fall off trees when winter approaches? If the leaves did not fall, the tree itself could not live throughout the winter months, for without the nourishment from the fallen leaves the tree would wilt and die. Why is it that trees (take a pine tree for example) have two types of pine cones? Their purpose for existing is to produce more trees until from one tree there emerges a forest. Why is it that a tree has ducts, the xylem and phloem? The ducts transport food and water throughout the tree. Why is it that trees have roots? To absorb moisture and soil nutrients from the earth, a tree must have roots. Why is it that a tree has bark? To protect the organism from parasites and inclement weather. Why does the tree itself exist? Trees provide shade for the thoughtful, comfort for the weary, inspiration for the human mind, and lumber to house the human body. A tree is not just a happening, no evolutionary theory can explain its origin, no scientific theory for that matter can explain its origin. The tree, as anything else in nature or life, is here for a purpose, a reason. A tree is a phenomenon like the earth and the stars, carefully constructed, carefully planned, on earth for a reason, on earth for a purpose.

"Change blows through the branches of our existence. It fortifies the roots on which we stand, infuses crimson experience with autumn hues, dismantles Winter's brittle leaves, and ushers Spring into our fertile environments. Seeds of evolution burst from their pod cocoons and teardrop buds blossom into Summer flowers. Change releases its redolent scent, attracting the buzz of honey bees and the adoration of discerning butterflies." *(B.G. Bowers)*

Light and Life

- Life is light and light is life.

- "What a seed lacks in interest, it makes up for in miracles." Life requires both light and water as does the seed in it's inchoate stage and throughout cultivation to fruition.

- Light is a source of life. We depend on it for night travel (artificial light) and in the day (sunlight). Everyone has their own switch that can be turned off or on voluntarily or involuntarily. Why are animals and insects mesmerized by light? Why is darkness so devoid of life? Types of lights: car lights, street lights, room lights, billboards, neon lights, fluorescent lights, incandescent lights, fireplaces, candles, laser, fireworks, flashlights, interior and exterior lights, strobe lights, spot lights.

- It maybe night outside but even in darkness there is eternal, ageless, light in the stars heavenward.

- Without light there is no color, no life.

- *"God is light" (1John 1:5)*

  Psalm 27:1 Psalm 27 Psalm 27:1-2 "The Lord is my light and my salvation, who shall I fear? The Lord is the strength of my life, of whom shall I be afraid?"

Genesis 1:3 Genesis 1 Genesis 1:2-4 "And God said, Let there be light, and there was light."

Ecclesiastes 11:7 Ecclesiastes 11 Ecclesiastes 11:6-8 "Truly the light is sweet, and a pleasant thing it is for the eyes to behold the sun."

Incidentally, I have to wonder what I would see if I traveled on a beam of light. According to Albert Einstein I would become light.

Genesis 1:16 Genesis 1 Genesis 1:15-17 "And God made two great lights, the greater light to rule the day, and the lesser light to rule the night: he made the stars also."

Water and Life

- Ninety-nine percent of the human body is water, the majority of earth is water, water evaporates, vaporizes.
- Water can be holy, water is in our blood and veins, water fills our oceans, rivers, and lakes, water extinguishes fire, we are weightless and dependent in water.
- Sometimes our memory is not clear as our vision is sometimes not clear, fuzzy, swimming through water, outer space. Frozen water can melt. There is fog and rain.

Baptism invites each of us into a new beginning. The images of water from Isaiah can help us reflect on this reality, through the waters of baptism the Lord "opens a way . . . and a path" for us (Isaiah 43:16).

> The powers of evil in our lives are "quenched like a wick" (Isaiah 43:17)
>
> "Then the Lord does 'something new' in us. The water 'spring forth' from within the wasteland of our hearts to create a river of new life. Little by little our dry deserts are transformed into lush gardens for the praise and honor of God." (Isaiah 43:19)
>
> Psalm 104:10-13 "He makes springs pour water into the ravines; it flows between the mountains. They give water

to all the beasts of the field; the wild donkeys quench their thirst. The birds of the air nest by the waters; they sing among the branches. He waters the mountains from his upper chambers; the earth is satisfied by the fruit of his work"

Jesus answered, *"Everyone who drinks this water will be thirsty again, but whoever drinks the water I give him will never thirst. Indeed, the water I give him will become in him a spring of water welling up to eternal life."* (John 4:13-14)

The Wind

Samuel 22:11 Samuel 22 2 Samuel 22:10-12 "And he rode upon a cherub, and did fly and he was seen upon the wings of the wind."

And while we cannot see the wind, it kisses our faces and rustles the leaves on our trees. It is akin to our Faith. We see the effects of the wind, we feel the effects of the wind, and we hear the wind so it sustains our Faith that it exists even though we don't see it.

Revelation 7:1 Revelation 7 Revelation 7:1-2, "And after these things I saw four angels standing on the four corners of the earth, holding the four winds of the earth, that the wind should not blow on the earth, nor on the sea, nor on any tree."

Time

- To a moth or butterfly, minutes, hours, days are like a life in years for us.

- Looking into the past > stars, light years away. "It's as if many of the stars we see glimmering in the evening died millions of years ago but their souls continue to live on as evidence by the unending light they caste." We can see into the future by seeing the past in the present.

- A movie/film projects images into the past (sights and sounds). We can see a whole life transpire before us on the movie screen, from infancy to middle age to old age, all in 2 to 4 hours.

- Time marches on

- Logic of creation is like clockwork, e.g. days, seasons, heartbeat, pulse, blood pressure, breathing, beats of the heart, batting of an eye

- Each day is historical

- Time marches on, but what is time? Time is space that knows no boundaries, clocks, segments, time lines. It's infinite.

- The fragility of life. What a difference 24 hours makes.

- Time is boundless space

- When we sleep at night a minute can seem as an hour, an hour as a minute, eight hours as seconds. There is no such thing as time. Man schedules events to try and make finite what is infinite.

- In fact film and pictures further exemplify man's futile attempts to make finite what is infinite; freeze time.

- We measure our perceived fit in time but time is infinity, time is endless, time is timeless

- We gauge our life by minutes, hours, days, weeks, months, and years but time takes no accounting because time is not quantifiable. What is a short day for one person can be the same for another or a long day by perception by yet another individual.

- Time is thought.

- Time is motion always moving forward.

- We can look back through history, pictures, film, the stars (light years to stars with only their image a fleeting vestige of their once viableness and reality).

- We can capture a dimension of a moment but moments are only memories and memories are only recollections, faded, some clear, and temporary.

- Time is as shapeable as putty or clay, a one or eight hour dream can be a lifetime experience, similarly a one hour movie or a four-hour movie can chronicle a lifetime.

- What is a moth's life is equivalent to a human life and what is a butterfly's life is equivalent to a human's and what is an animal's or fish's life is the same to a human.

- That butterfly's life of a few days or weeks could be decades to a human.

- We can compress 100 years into one or flashback a whole life in the recollection of one's last time on earth when dying.

- Time is something that consistently chases us, moving us, prodding us, pressuring us forward.

- There is no such thing as lost time, just the opportunity to sit on the moving side walk, sit on the speeding train, sit in the airplane or car headed to the next arrival and departing location.

- We cannot buy time, we cannot control or capture time, and we can only witness its power in controlling us.

- We can only travel with Time until it's ready to kick us off the ride.

- Time cycles include tide, ocean, seasons, weather, moving forward.

- Time is a Roller coaster, clock, earth all moving around.

- Sun up or sunrise; sundown or sunset

- Life and time is emotions, mood swings, ups and downs.

- Pictures allow us to capture or freeze the present which becomes the past, but we still can capture the future because it hasn't happened yet.

- Moment by moment. Life is like pages of a book, you don't know how next page reads, you read it line by line, page by page, day by day, insight by insight, mile by mile, minute by minute, wave by wave, frame by frame.

- The calendar is like a circle, a clock, a continuous cycle, the spokes of a wheel.

- During World War II when Nagasaki and Hiroshima were bombed the images of persons were flashed on sides of buildings or on sidewalks when atomic bombs were dropped.

- In Pompeii where victims of the volcano died their images were frozen by ash and fire in time.

- Observe the preserved bodies in ice caps and frozen areas of the last Ice Age

- Bible quote: "One day as a 100, 100 days as one."

- Think about it: When you are 1 year of age, turning 2 is equivalent to one-half of your entire life; When 62 years of age, turning 63 is only one percent of your life.

- Helen Steiner Rice wrote, "God never comes too early and He never comes too late."

- When people are dying or face death, life flashes before them in a split second and they vicariously relive childhood to adulthood, their most important life experiences.

- Wonder what I would see if I traveled on a beam of light. According to the scientist Albert Einstein if I traveled at the speed of light I would become light.

- We, as a human race, share history/experience with our ancestors by viewing the same moon, planets, stars as our forefathers saw thousands, millions of years ago.

- The mind never sleeps, the body does.

- "We must always have old memories and young hopes."

- The Wedding ring has no beginning and no end. We as humans interrupt the cycle with adultery and divorce.

Seeing or Not Seeing Is Believing

- "Blessed be the blind man, for you shall see with new eyes."
- We can't see our backs. We never really see all of ourselves in a lifetime, why? Why do we generally face forward? One of life's mysteries.
- "The farther backward you can look, the farther forward you can see." (Winston Churchill)
- Science is creationism, and vice versa, they go hand in hand. Each explains and illustrates the other. They are partners in Life.
- "Blessed are they who do not see but believe."
- "Once I was blind but now I can see." (19 John, Chapter 9)
- Regeneration of the flying seed. From Dust to Dust.
- We don't see our fate, disease, the chemical changes in our bodies, historical evolution, we cannot see microscopic forms of life yet all the aforementioned Exists.
- All faiths and religious denominations appear to share a common Bond/Believe: All Believers Believe in God, The Creator.
- People say that because they have never actually seen God that God does not exist. They are partially correct when they say

they have never seen Him. Neither have I, but I have seen evidence that reveals His existence. It is a great fallacy of people to say that something does not exist just because they don't actually see that something. Frankly we did not know that certain planets existed until they were visually discovered via the telescope. The fact is that just because one does not actually see something does not mean that that something does not exist. Similarly we cannot see a brain or mind think, but we know it is thinking or at least has the capacity thereof. We know that conscience exists—a guilt feeling we possess when we've done something wrong—yet it is not tangible; we cannot see conscience. Yet, are we to deny its existence? Are we to deny that thinking exists? We cannot see that entity which we identify as thinking.

- And what about microscopic life? We cannot see tiny life forms except with the help of a microscope. So microscopic life is there, it exists, but we can't readily see it with the naked eye.

- The higher up in space/altitude that you find yourself, the smaller are things below, e.g. geography, countries, planets. You don't even see people.

  "The hearing ear and the seeing eye, The LORD has made both of them." (PR 20:12)

  "And He said, 'to you it has been granted to know the mysteries of the kingdom of God, but to the rest {it is} in parables, so that SEEING THEY MAY NOT SEE, AND HEARING THEY MAY NOT UNDERSTAND." [Lu 8:10)

- An unknown author once said, to quote, "A fine artist paints what he sees; a folk artist paints what he knows is there."

Colossians 1:16,17 – "For by him were all things created, that are in heaven, and that are in earth, visible and invisible, whether they be thrones, or dominions, or principalities, or powers: all things were created by him, and for him: And he is before all things, and by him all things consist."

Hebrews 11:3 "By faith we understand that the universe was formed at God's command, so that what is seen was not made out of what was visible."

Job 12:7-10 "But ask the animals, and they will teach you, or the birds of the air, and they will tell you; or speak to the earth, and it will teach you, or let the fish of the sea inform you. Which of all these does not know that the hand of the LORD has done this? In his hand is the life of every creature and the breath of all mankind."

Psalm 19:1-6 "The heavens declare the glory of God; the skies proclaim the work of his hands. Day after day they pour forth speech; night after night they display knowledge. There is no speech or language where their voice is not heard. Their voice goes out into all the earth, their words to the ends of the world."

Romans 1:19-20 "What may be known about God is plain to them, because God has made it plain to them. For since the creation of the world God's invisible qualities—his eternal power and divine nature--have been clearly seen, being understood from what has been made, so that men are without excuse."

- **Everything in nature does what it is designed to do. It may appear as though nature operates like a well-built clock, but**

nature is not the marvel. It is merely obedient to God. God is the marvel behind nature.

John Roberts in his Essay's, says, "God has made nature to be vaster and more complex than anyone is capable of comprehending. What He has revealed to us about nature is not accidental. He has chosen what we are to understand about nature. He has also chosen 'the secret things' that belong to Him, which will not be revealed."

The Father and The Son Proclaiming the Gospel

- Jesus, through deed, words, stories, healing, exorcism, mingled, ate with disciples, taught in parables, lived paradoxically.

- Look without seeing, listen without understanding

  Matthew 13-25, "Talk in parables. Parables shocked surprised, reversed expectations of common people and their assumptions of good and evil."

Jesus taught three things about the Kingdom:

1) The Kingdom is near (coming soon)
2) The Kingdom is personal (healing)
3) The Kingdom comes through vulnerability (violent people destroy it, but the kingdom does not come by force because the people change their attitudes and behavior by choice)

The Sky
One sky past and present
One sky umbrella over all of us
One sky ours forever
One sky always upward

- "The hearing ear and the seeing eye, The LORD has made both of them." (PR 20:12)

- "And He said, 'to you it has been granted to know the mysteries of the kingdom of God, but to the rest {it is} in parables, so that SEEING THEY MAY NOT SEE, AND HEARING THEY MAY NOT UNDERSTAND." [Lu 8:10)

- Ultimately all of the ways that nature is used by God can be combined into one major overall purpose for nature: Nature is used by God to focus our attention and worship towards Him.

The Image and Likeness of God's World

"Poor indeed is the man whose heart is not warmed, whose soul is not inspired, whose mind is not enriched by the indescribable colors of which The Creator has painted the birds and flowers of His world. To every person so moved there is always the yearning to copy the beauties of nature yet how few are endowed by the Master Artists with that ability." [Source: Judge James Lindsay Almond, Jr., US Court of Customs & Patents Appeals, Washington DC]

Behold the Kingdom

"The multitude followed a man—a prophet who spoke words of wisdom. And they listened, trying to understand the paradox of his great Truth.

He said, "Blessed be those who are poor, for you shall inherit the Kingdom and blessed be those who are weak for you shall inherit great strength, and blessed be those who are children, for you shall be counted as wise, and blessed be the blind man, for you shall see with new eyes."* (Behold the Kingdom From The Painter by John Michael Talbot)

> Mark 4:30-32 - He also said, "With what can we compare the kingdom of God, or what parable will we use for it? It is like **a mustard seed**, which, when sown upon the ground, is the smallest of all the seeds on earth; yet when it is sown it grows up and becomes the greatest of all shrubs, and puts forth large branches, so that the birds of the air can make nests in its shade."

Reflection

I've been, as many of my brothers and sisters, on commercial airline flights, in the cockpit of a small engine plane, a helicopter, or in a hot air balloon and can relate to the quiet, removed, surreal feeling one gets atop the world and temporarily aloof from its concerns and daily pace. This sterile, tranquil, environment is as peaceful as the sound of ocean waves crashing against the shore and as sobering as the withdrawal of the sun from the shoreline as it descends out of our sight. We deliberate on how alone we are, at times, without the support and love of our Lord and family. We pause on how insignificant we are in proportion to the larger whole: one grain of sand in a sea of sand. It makes one yearn for an escape from this abeyance, to land safely on the runway, and caress the warmth of the ground with our feet and hands from which we came and to which we will return: dust to dust.

- Regeneration cycle, e.g. examples include scarlet salmon run (birth cycle to death), tad pole regeneration, sharks teeth, fish returning to their place of birth to die but also to give life to next generation. Crabs leave their shells, their bodies behind. The caterpillar becomes a butterfly.

    Psalm 145:3-7 "Great is the LORD and most worthy of praise; his greatness no one can fathom. One generation will commend your works to another; they will tell of your mighty acts. They will speak of the glorious splendor of your majesty, and I will meditate on your wonderful works. They will tell of the power of your awesome works, and I will proclaim your great deeds. They will celebrate your abundant goodness and joyfully sing of your righteousness."

Blind Faith

- We have Faith that sun will rise or fall
- We have Faith that no deep freeze or ice age will occur
- We have Faith that the earth will not be struck by a meteor from outer space
- When we sleep we depart from reality, we go into the realm of the unknown and have Faith that we will wake again to see another morning.
- From the movie "Miracle on 34th Street", to quote, "Faith is believing even though common sense tells us not to."
- Throughout God's creation there are many things that we recognize and accept without a passing thought – the light of day, the air we breathe, and the presence of gravity. These things are obviously part of what we call nature. A less obvious aspect of nature is the absence of things. The absence of water is dry. Deserts are places best defined by the absence of water. The absence of light is dark, as in the dark of night or the darkness of shadow or shade. The absence of wind is called calm or perhaps quiet which is also the absence of noise. Whether it is the desert or the tropics, the night or the day, the howl of wind or the dead calm there is no question that such things are a normal part of what we know as nature.

Everything Lives In Its Own World

Everything lives in its own world, universe, most of the time oblivious to our world and our interdependence. The bird, the squirrel, the chipmunk, fish, wildlife, insects live independently and focused on their own survival, milling about their own microcosms, co-existing with us yet literally totally distinct. We see them, they see us sometimes, but we go our separate ways.

We all live in our own little worlds. Have you ever thought about it? While we co-exist with other humans in life we nevertheless fill our own shoes and stand alone—despite the love of family and friends—in a sea of humanity. We are responsible for our own actions, conduct, behavior, attitude, likes and dislikes. Our destiny is, in part, left to opportunity, luck (where time meets opportunity), chance, and interaction with others, dependence on others, our own choices and decisions. Our faith and belief in God.

The bottom-line, however, is that we face our own destiny, present, past, and future. The amazing thing is that we exist in concert with other creatures, or living beings, that simultaneously live and breadth within their own, or our, ecological niche. Fish swim in the sea, the ocean, a lake, in ponds, and in rivers but oblivious to our existence except perhaps when a fisherman fishes which negatively impacts their lives. Even fish who swim in an aquarium live in their own world. We live in coexistence with animals except when hunters shoot them and negatively impact their lives. Even animals who live in zoos are blind to our existence for the most part. We live in harmony with birds and butterflies for the majority of the time, even insects, except when the latter became annoying pests. We live surrounded by forests, flowers, trees, bushes, and plants albeit we many times use them to support our lifestyle with gardens, adorning our yards, mowing our grass, pruning trees, trimming shrubs, or cutting trees for home building, fire wood, or construction projects, using these natural things as our resources. However, in the majority of cases, we live separate and apart—albeit simultaneously—with flying, crawling, walking, beings replete within our environment, i.e. ants on the ground, birds in the air, amphibians and reptiles in the water, a plethora of creatures great and small sharing our common space. It's mind boggling when you consider that we can't put ourselves into the minds and bodies of said creatures to know what they are thinking, the reasons for their actions, their mission in life, however mundane or simple. In most cases the lower beings, similar to ants, flying insects, birds, and animals go about their business as God intended them to and except for our interference live their reason for being on earth.

The World Isn't Just an Evolutionary Happening Or Result of *The Big Bang* Theory

When you observe how a clock or watch is made you notice the little intricacies of its construction. The way one small appendage is attached to another, a spring attached to a coil, a coil to a spring. You notice the precise order, the delicate balance of one part onto another. If one observed a leaf from a tree or a flower he undoubtedly would note how each part of the leaf or flower was put together. The pedals surround the pistol and the pistol, in turn, is surrounded by filaments. The precise order and balance that is seen in man-made objects can also be observed in nature.

Order is something that is made out of something unorganized—order is organization, precision, balance. Purpose is the reason for why order is made—the reason for making something organized out of something disorganized. Order and purpose are products of the mind and can only be the products of mind, for they are also the products of reason, and reason is a product of the mind.

God used His mind to make an orderly world and he made everything on earth for a purpose. People must grasp the idea that everything on earth and the earth itself is not just an evolutionary happening but a result of the careful planning and orderly thinking of a Great Mind—fathoms greater in scope than ours. Happenings do occur, but not in any particular order. Happenings such as mutations do occur, but have no purpose behind that occurrence; they just happen. The world isn't just a happening; God made it. God Exists.

Seeing with all our senses

In this time of turbulence, with world tensions, terrorism and the unpredictable economy affecting us all, it is important that we not let negative environmental factors take control of our lives. Rather, we should focus on the big picture, which takes in the silver lining of every gray cloud. It is imperative that we not take for granted our surroundings, those daily occurrences in nature such as the change in seasons, the blazing color of autumn leaves, the falling manna of snowflakes in winter, the warm sun against our cheeks in summer and the glorious array of aromatic blossoms in the spring. Just as we daily take notice of the ever-changing patterns of weather—from cooling rain showers to the warming sun, from hot summers to cold winters—we must learn to appreciate everything in nature that conjures up emotion, from nostalgia to elation.

There is a bright aura circling Earth, tenuously separating our planet from the dark recesses of outer space. While watching an IMAX film at Cape Kennedy in Florida, I was transfixed by this border between life and death, as seen from satellite reconnaissance in space. It appeared like a halo surrounding Earth's circle, a vulnerable umbrella that, if ever expunged, would expose us to outer darkness and the absence of life. The movie expounded on how precarious our life on Earth has become, with the imminent threats of global warming, the deterioration of the ozone layer, the pollution of our water, air, and other pristine areas, and the negative influence of uncontrolled emissions from industry and automobiles. It was a sobering thought that compelled me to write this book.

In this book I articulate Albert Einstein's theory that stars are our looking glass into the past because they reached their demise millions of light-years ago and are only images of their prior existence. I examine the miracles of nature whereby the caterpillar metamorphoses into the butterfly. I discuss how the brief lifespan of an insect is, to it, tantamount to our human years, and that in an eight-hour dream, as we sleep, we can experience a lifetime, just as in a two-hour movie we can vicariously experience the passing of life's adventures over decades. I address how we live and die within our life cycles analogous to the seasons – with spring, as life's beginning, summer as life's prime, fall as a time for reflection and winter as life's end. It is akin to infancy, adolescence, young adulthood, middle age and old age being separate lives. In each case we live that phase and then are reborn into another chapter of life.

So we must witness the plethora of nature's miracles—the birds, flowers, trees, storms, lightening, rain, snow, animal life, fire, earthquakes, volcanic eruptions and other natural phenomena—regardless of our beliefs or non-beliefs, religious denominations or agnosticism.

I became strongly aware that there was a power beyond any of us when I witnessed an autistic child, blind and deaf, portrayed on a television documentary one evening. He played piano selections from Mozart and Chopin without ever having seen the musical scores before, and I cried. Seeing insects and lizards dancing across the water's surface also has convinced me that nothing is impossible. We all must recognize that there is something operating in this world that is greater than us, something to be revered and marveled—and something that should give us the faith to sustain our everyday lives. We don't always have tangible evidence, for seeing is not necessarily believing. We cannot see microscopic life, but it is nevertheless there. And we cannot see the wind, but it kisses our faces and rustles the leaves on our trees.

> Song of Solomon 2:12 Song of Solomon 2 Song of Solomon 2:11-13 "The flowers appear on the earth; the time of the singing of birds is come, and the voice of the turtle is heard in our land."

God's Rainbow

His Holy Covenant connecting Humanity with Nature and all Life on Earth

Genesis 9:8-10, 12-13, 16-17

Then God said to Noah and to his sons with him: "I now establish my covenant with you and with your descendants after you and with every living creature that was with you--the birds, the livestock and all the wild animals, all those that came out of the ark with you--every living creature on earth And God said, 'This is the sign of the covenant I am making between me and you and every living creature with you, a covenant for all generations to come: I have set my rainbow in the clouds, and it will be the sign of the covenant between me and the earth.... Whenever the rainbow appears in the clouds, I will see it and remember the everlasting covenant between God and all living creatures of every kind on the earth.' So God said to Noah, 'This is the sign of the covenant I have established between me and all life on the earth.'"

We Ask Why

John Roberts from his Essays elaborates, "In praise of God – it is no mistake that God has placed us in nature, and placed nature inescapably, around us, yet He has limited our ability to comprehend the fullness of where He has placed us. This is by design. Nature is his creation just as we are his creation. That He has both endowed us with the desire to explore the question "why" and placed us in a world where the answer is beyond our comprehension is not by mistake. It is a gift from God, the purpose of which He has not hidden from us."

Philippians 4:8 suggests, that which God creates is worthy of special attention, worthwhile for us to pay attention to. The song of the loon, the colors of an autumn leaf, the reflection of moonlight dancing on water, a dewdrop laden spider's web – the pure, pointless beauty of things small and grand – why should any of this be noteworthy or important to us? Why do the scriptures, God's written word, suggest that they are?

He remarks, "In a sense, God has provided us a picture book that helps us to understand his printed Word. When you open your Bible you don't immediately see pictures. Look more closely, though, and you will find words like lightning, fire, wind, seeds, flowers, trees, and birds embedded everywhere in its passages. These are all things of nature – common things, familiar things. God created nature to be all around us. To live on earth means that you are immersed in nature."

Roberts explains, "Rock and light – these are more than words in a sentence. They are power-packed pictures on a page. Because of their connection to creation and our familiarity with the things in nature, what we can comprehend is not just light or a rock. From these words we are better able to understand the complexity of spiritual things and the hidden things in God's Word."

Going Where the Bible Isn't

To quote Roberts, "The Bible, as a pathway to understanding God, is not available to everyone, 'millions of people are living their lives with no access to the Scriptures.' However, from what has been made – nature, God provides a pathway to understanding, that is available to anyone. Poets, artists, musicians, writers, through all time, have recognized and proclaimed that from the experience of nature comes recognition of God."

> *"Nature is a greater and more perfect art, the art of God."*
> Henry Thoreau

> *"Bless Thee, O Lord, for the living arc of the sky over me this morning."* Carl Sandburg

He concludes, ". . . If you recognize nature as God's Creation, trust that God has placed us in nature for his purposes, know that everything in nature points you towards God . . . you will gain spiritual insight. As it is written in Romans 1:20, the invisible qualities of God—his eternal power and divine nature—will begin to be seen and understood more clearly."

**(Note: John Roberts is the author of an Essay entitled God's Wonder Through Nature: A Picture Book)

Creationism Versus Evolution

At a time when geologists dismissed the idea of a global flood and recognized Earth's age, many theologians acknowledged that there was more to the past than articulated in Genesis. Conservative creationists rejected this perspective and chose to see geology as a challenge to their faith.

Faith in Nature

For the first millennium of Christianity, theologians embraced knowledge of the natural world in order to thwart challenges to biblical authority. Saint Augustine, Thomas Aquinas, and John Calvin all adopted reason as the way to learn about the world. Augustine was among the first to guard against promoting biblical interpretations that conflicted with what one could observe for oneself. Centuries later, Aquinas praised the pursuit of knowledge gained from experience reading God's other book—nature.

At the time of the Reformation, Calvin also considered the revelations of both nature and the Bible as fundamental truths. In his *Institutes of the Christian Religion*, Calvin supported the idea of respecting natural truths revealed through the observation of nature: Calvin believed in keeping an open mind when it came to what we can learn about the natural world from observation. From his perspective, closing one's eyes to the way the world works was to close one's eyes to God.

Augustine, Aquinas, and Calvin all believed that Noah's Flood was a global flood. They shared the consensus that fossil seashells found in rocks were compelling evidence—how else could the fossils of marine creatures become entombed in rocks high in the mountains?

The Roots of Creationism

The roots of modern creationism can be attributed back to George McCready Price, an amateur geologist. In a treatise that appeared to be a geology textbook, Price contended that there was no order to the fossil record. He asserted, instead, that the succession of organisms which geologists read into the fossil record was really just a potpourri of samplings which thrived in different parts of the antediluvian world.

Leading fundamentalists praised Price's book, calling it a monumental work and a most creditable presentation of the Science of Geology from the standpoint of Creation and the Deluge.

The preponderance of Christians continued to support attempts to reconcile geology and Genesis. Twentieth-century fundamentalists split into young-Earth creationists who defended a global flood, and old-Earth creationists who acquiesced that there was geological evidence that we live on an ancient planet.

Creationism Today

"Even though debates over the geological implications of biblical interpretations are historical to the earliest days of the Church, the story of how naturalists struggled with reconciling the flood with a growing body of contradictory geological evidence shows that the twentieth-century revival of flood geology recycled ideas previously abandoned in the face of challenging scientific evidence. In view of nineteenth-century scientific discoveries, it appeared reasonable to read the biblical account of the Flood as allegorical. Consistently through time, Christians were neutral to geologic findings by reinterpreting Genesis to preserve the integrity of both natural and scriptural truths." *

*Source: David R. Montgomery, The evolution of creationism, *GSA Today*, v. 22, no. 11, p. 4-9, 2012. With express permission from the author.

==

My rejoinder to the testimony of rocks and evolutionary geological history

My rejoinder to the evolutionists and academicians, who argue geological evidence as proof discounting a global flood, or second-guess the creation of earth in six days, is the outcome of narrow minded "scholars" who sit in ivory towers and whose focus is not objective, or open-minded, but limited to the historical record of rocks. These "rock heads" "do not see the trees for the forest". They do not broaden their perspective to the more compelling natural evidence, aside from geological theories, to include the following stalwart pillars of thought which, if one uses common sense, versus so-called scientific knowledge, refute any and all contentions contrary to traditional Faith. Namely,

- The earth and its inhabitants reflect the work of a Higher Being

The *cell* is the building block of life. It can only be viewed with a microscope, yet it is more complex than the intricate workings of any computer. If computers are made by intelligent humans, why not the cell as a creation of a supreme being?

Inside the cell are the *genes* that determine heredity and the DNA footprint. The information in the genes of one cell would require a thousand-volume encyclopedia. Encyclopedias are written by smart humans. Can we doubt that the cell was the product of a supreme being?

The *eye* cannot be matched by any man-made camera. Cameras are made by intelligent beings with sophisticated technical minds. Who can believe the eye came from any source other than our Creator?

- Life is inter-dependent

Most living organisms in nature cannot exist without one another. Trees, for instance, reproduce sexually. Pollen contains sperm and can be transferred by insect or wind to another tree. Seeds are fertilized and the seeds grow into new trees. This is true of trees such as pine, apple, coconut, and many others. Cones, flowers, fruit, seeds, nuts, and pollen are all part of sexual reproduction. Some trees will multiply asexually. Aspen trees will send up runners from their roots which can grow into a whole separate tree. Many trees have both male (pollen-bearing) and female (fruit/nut bearing) parts in the same tree, but some, such as the ginkgo tree, are only one gender, though sometimes it can switch.

Flowers reproduce when pollen from the male stamen fertilizes the female organ called the carpel. All flowering plants have both male and female reproductive organs. The stamen of a flower is made up of two parts: The filament and the anther. Pollen grains are found on the anther, which is at the end of the filament. This stamen is located inside the flower's corolla, formed by the petals. The carpel, or female reproductive organ, is located in the center of the corolla. The carpel is comprised of three separate parts: The stigma, style, and ovary. Pollen from the male stamen fertilizes the female carpel.

Human and animal sexual reproduction are additional illustrations of where neither the male nor the female alone can reproduce without dependency on the other partner.

Another example is Photosynthesis or the process used by plants, algae and certain bacteria to harness energy from sunlight into chemical energy. During photosynthesis, light energy transfers electrons from water to carbon dioxide which produces carbohydrates. In this transfer, the carbon dioxide is "reduced," or receives electrons, and the water becomes "oxidized," or loses electrons. Ultimately, oxygen is produced along with carbohydrates. Evolution cannot explain this inter-dependence, but creation can.

In each of these aforementioned instances if one form of life developed gradually over millions of years, it would have reached its demise because it could not reproduce or exist without the other kind.

Throughout the course of our historical legacy of countless years, both living entities would have to develop simultaneously and then discover one another before they expired. Otherwise they could never live or reproduce.

Creation Proves God's Unlimited Wisdom

No human wisdom or other material wisdom is capable of creating the universe. Creation proves that God not only exists but also is infinitely wise.

> Jeremiah 51:15 - He has established the world by His wisdom, and stretched out the heaven by His understanding. [10:12]
>
> Proverbs 3:19-26 - The Lord by wisdom founded the earth; by understanding He established the heavens.
>
> Ecclesiastes 11:5 - As you do not know the way of the wind, or how the bones grow in the womb of her who is with child, so you do not know the works of God who makes everything.
>
> Romans 11:33-36 - All things are of Him and through Him (i.e., created by Him) and unto Him. His ways and judgments are unsearchable and past finding out. Who can know His mind or give Him counsel?

The Biblical Approach to the Evidence of Creationism

The Bible instructs us on how to approach the evidence of Creationism. This approach is based on observing nature, and does not mandate scientific knowledge.

Romans 1:20-25

God's invisible attributes (Deity and power) can be seen and understood by the things He made. So we need not be scientists, capable of arguing scientific technicalities, to settle the issue of God's existence. The evidence is 'clearly seen' by observing the universe. There is no excuse for those who do not recognize this; so this is something any of us can see for ourselves.

Psalms 19:1-4

The heavens declare God's glory and the firmament shows that it is the work of His hands. We can learn this as we observe the heavenly bodies, such as the cycle of day and night. The message is not spoken in human language, yet it has gone to the whole world. Again our own observation can convince us that the universe was made by the hands of God.

Job 12:7-10

The beasts of the earth, the birds of the air, the fish of the sea, and the earth itself likewise explain and instruct us. Who among them does not know that God's hand did all this? Of course, they do not know it consciously. The point, as with the heavens, is that their nature and existence proclaim this message to us. Observing them should convincingly demonstrate the truth, so these are lessons we can learn from them.

Psalms 139:13,14
We should praise God, because we are fearfully and wonderfully made. God formed us, His works are marvelous, and we should know that very well. So the human body also serves to convince us that God is creator.

All the above Bible passages in *The Biblical Approach To The Evidence of Creation* show that, even without technical knowledge, we can observe the earth around us and conclude that there must be a God who made it.

Common sense allows us to conclude that the amazing attributes of plants, animals, and people were all logically made by a supremely intelligent being. Evolution contends that all this happened by chance – the Big Bang Theory - or that life commenced by spontaneous combustion, evolving via random mutations. There is simply no scientific evidence that life can suddenly appear where there was no life before.

However, it does not mean that science contradicts the Bible view of creation. Instead the better we understand science, the more data we discover to confirm creation.

Actually, both evolution and creation are matters of ***faith***, not ***personal observation.*** Faith involves believing a thing on the basis of evidence, though we never personally saw it. In the case of origins, the evidence consists of inferences drawn from scientific evidence. But in addition, creation also has the eyewitness testimony of the Creator who inspired many Bible writers to record what He did. This harmonizes with creation, which says that all life was created by an eternally living God. "It is shameful to say man is the offspring of a stone (Jeremiah 2:26,27)."*

*Gospel Way, A Biblical Approach to Evidences for God, Jesus, and the Bible. With Express Permission from Reverend David Pratte.

Three Major Religions, One God

"Three of the world's major religions – the monotheist traditions of Judaism, Christianity, and Islam – were all born in the Middle East and are all inextricably linked to one another. Christianity was born from within the Jewish tradition, and Islam developed from both Christianity and Judaism. While there have been differences among these religions, there was a rich cultural interchange between Jews, Christians, and Muslims that took place in Islamic Spain and other places over centuries"*

*The *Global Connections* Web site integrates and contextualizes the rich body of ***public broadcasting*** resources to provide a global and historical perspective that will help teachers, students, and the general public explore and understand seminal events of national and international significance.

Universal Love

I believe in the concept of ***universal love***, the idea that we, as people of the world, all have blue veins and red blood despite maybe looking different, of disparate cultures, speaking a myriad of languages, and having our own religious or secular perceptions of reality or our Creator.

We are no doubt a diverse world with a plethora of opinions, thoughts, and beliefs just as there are distinct fingerprints, foot prints, unique DNA, minds, hearts, and souls.

We are of different races, e.g. red, white, brown, black, and yellow, as well as genealogy, ethnicity, and history.

Yet we all share the same globe, common space on the planet earth.

We experience the same rain, wind, snow, warmth, hot and cold temperatures, weather, climates, and changes in the environment.

We look upon the sky, the stars, the clouds, the universe, outer space as one.

We cannot escape the reality that the world is not flat but round with no beginning and no end, a circle, if you will.

We benefit from the same natural resources, share and travel the same rivers, oceans, lakes, mountains, valleys, desserts, forests, and other natural wonders.

We have to conclude that nature and science, or Creation, is our common ground, the link connecting us all, our way of understanding and communicating with each other, God's Covenant with all peoples.

We are all bound by and dependent on nature just as we are by God's Covenant with us that requires our reverence, dedication, commitment, allegiance, and Faith to abide by God's laws and rules for living in nature.

We all derived from Adam and Eve and the Garden of Eden.

Regardless of whether we are Hindu, Buddhist, Christian, Judaic, Protestant, Confucianism, or whatever religion, we are all created by God – however your Creator is perceived according to your religious convictions – and we all have hearts, minds, and souls.

We all have the capacity to love and care about others, to relate to and get along with people of other backgrounds, and we all have choices available to us and a conscious.

While there are disagreements and conflict between countries and people, this is not inconsistent with co-habitants in oceans, rivers, lakes, forests, or the world's ecological niche.

The bottom-line is that there are wars and conflict between peoples, differences of religious principles, beliefs, and opinions but this is consistent with life on earth where birds and animals compete for food and living space, or where that same competition is characterized as a "survival of the fitness" in a natural environment of predators and the hunted.

With God's exceptions, people are for the most part created in God's image with the capacity to see, hear, touch, feel, and smell nature. Given many persons have disabilities; nevertheless people are in the majority of instances made with a head, a body, and arms and legs.

We all share the Creator's prototype in other words. We are all constructed from the same mold or template.

We share the same sights, sounds, smells, and amazement of trees, plants, flowers, birds, animals, water, light, thunder, storms, and wind.

The thread that binds or sews us together and forms a common bond is nature, or God's means of uniting us as one body.

We are all Creatures of nature, of science, of God. As such we all adore, revere, and worship that which God created—his nature, his wonder—and while we contend that we believe differently, we are mostly aligned, or in sync, with our Faith than we realize.

There was a PBS documentary which aired a number of years ago and delved into the various religious denominations with the finding, the conclusion if you will, that we basically, with minor exceptions, believe in one God, the maker of Heaven and Earth.

We are all bound by God's commandments defining the parameters of our moral conduct and/or sacred agreement to adhere to God's wishes for the code of our conduct so that we may inherit the Heaven he made as an extension of our Earth.

A common theme interconnecting all religions is the moral requirement that we do right in God's eyes and therefore keep our commitment to God's covenant.

We share the Holy Spirit which we may not see but all feel and experience just as we feel the wind on our faces or against our flesh.

A brief history of Christianity

Christianity started as an offshoot of Judaism in the first century C.E. Until the emperor Constantine converted to Christianity in 324 C.E., early Christian communities were often persecuted. It was then that the Roman Empire became the Holy Roman Empire, and its capital relocated from Rome to Constantinople (formerly Byzantium and now Istanbul). The development of Christian groups derived from major and minor splits.

The Orthodox Church and its patriarch split away from the Roman Catholic Church and the Pope in 1054 C.E. because of political and doctrinal differences. In the 16th century, Martin Luther, upset at the corruption of the Catholic papacy, spearheaded a reformation movement that led to the development of Protestantism.

Christian missionaries proselytize all over the world, and there are large populations of Christians on every continent on Earth, although the forms of Christianity practiced vary.

Creationism, Science and The Catholic Church

The Catholic Church does not appear to reject science, nor does it subscribe to contentions of the scientific community. Instead, the Church rightly commends science when it does something significant for humanity such as seeking a remedy for a life-threatening disease. By the same token, the Church opposes scientists who violate fundamental moral principles on such things as embryonic stem cell experimentation.

The question that the Catholic Church asks when examining a scientific theory is does it contradict Scripture? which is regarded as infallible. Any Catholic theory must adhere to the following Dogmas:

1. God created everything out of nothing
2. The universe is not a product of chance
3. Everything depends on God for existence

The age of the earth is not about Evolution

Darwin was a biologist and his theories are biological. Evolution is not a fact, but a set of theories. Some of the theories are compelling, such as fossil records and observed micro evolution within species, while other aspects of evolutionary theory have been proved wrong by science itself. The complexity of each species is something modern science has been unable to explain. The age of the earth is derived from distinct areas of astrophysics.

Debating Creationism

Believers and non-believers get agitated about the subject of Creationism, almost as if our salvation depends on it. Some Christians attribute modern problems to society's lack of belief in Creationism. Other Christians say that the insistence to preach Creationism alienates many persons from Christianity.

Some Catholics believe that God created the universe verbatim as laid out in Genesis 1 - a young earth. Other Catholics believe in an old earth. The Church seemingly has no defined Dogma regarding the specifics of how the earth and the human body were created.

Interpretation of Scripture

There are various ways to articulate truth, one being through scientific language. Another is through literary means. Some of our greatest truths are communicated using literary prose.

When interpreting Scripture and what it is conveying to humanity, the reader needs to focus on the genre that is being presented. Some sections of the Bible are historical facts, some are allegories, others are poetic. For instance, the Gospel of Luke describes events from eyewitness accounts and therefore is historical. As for books like Job and Jonah, the Catholic Church is open-minded.

Did God create the world in 7 days?

It would be safe to assert that most Catholics conceive the world as ancient. They do not think this interpretation is liberal or a compromise of God's Word. They would say that in the book of Genesis, God was using the language of passion rather in place of science. They emphatically assert that everything in creation happened as described, and that it not just a story.

> "a thousand years in your sight are like yesterday when its past or like a watch in the night" (Ps 90:4)

If there ***was*** a "Big Bang" and a certain amount of progression and change within each species, it was God that made it happen.

The Creation of the Human Body

The Church does not have an official teaching on the origin of the human body. There are several religious perspectives which are not contrary to Catholic theology.

1. God directly created human beings.
2. God designs the laws of the universe, so that they will produce the human body through natural processes.
3. God designs the laws of the universe and intervenes directly in history to create life in general.

Creationism or evolution respond to two different realities. The story of the dust of the earth and the breath of God does not in fact explain how human persons come to be but rather what they are. It explains their inmost origin and casts light on the project that they are. In contrast, the theory of evolution seeks to understand and describe biological developments. But in so doing it cannot explain where the 'project' of human persons comes from, their inner origin, nor their particular nature. To that extent we are faced here with two complementary -- rather than mutually exclusive -- realities.

The Creation of the Human Soul

On the creation of the soul, the Church has a very strong teaching. The human soul was purposely created in the likeness and image of God.

- The human soul is not simply a byproduct of the human body.
- The human soul has the power to know abstract concepts, to know God.
- The human soul has the power to choose and to love.
- The human soul has unique dignity above the rest of visible creation.

The human being is a combination of human body and human soul. Regardless of any speculative ideas of evolutionary processes that God may or may not have used in the design of the human body, Adam and Eve became human beings when God infused their bodies with human souls. The creation of the human soul was created immediately.

Science should be the advocate of Christianity, not the adversary

When science is applied as a pure art, it is neither atheistic or pro-religion. It simply tries to ascertain the truth based on natural observation of phenomenon. However, what has happened in the last century is that human secularism has sought to hijack science for its own purposes.

> 159. Faith and science: "...methodical research in all branches of knowledge, provided it is carried out in a truly scientific manner and does not override moral laws, can never conflict with the faith, because the things of the world and the things of faith derive from the same God. The humble and persevering investigator of the secrets of nature is being led, as it were, by the hand of God in spite of himself, for it is God, the conserver of all things, who made them what they are." [Vatican II GS 36:1]
>
> The Church does not propose that science should become religion or religion science. (JPII)
>
> The unprecedented opportunity we have today is for a common interactive relationship in which each discipline retains its integrity, and yet it's open to the discoveries and insights of the other. (JPII)

Science cannot prove or disprove God's existence because God is outside the limits of empirical measurement.*

* Catholic Bridge.com

Some Concluding Thoughts

So you may have read this book hoping to maybe come to some conclusion about the existence of your Creator, God, The Supreme Being, The Holy Spirit, only to derive at more questions to which none of us, including this author, presume to have the answers. Instead, however, of ending our journey here I hope that this adventure in thought has stimulated you, motivated you if you will, to make inquiries of your own and to seek answers as you move forward in time. Don't just sit idly by and refuse to dedicate yourself to attempting to understand why you are here, your purpose in life, and what you can do to make your being, and the lives of others more worthwhile. We are all here for a reason, we all have a mission in life even if ill-defined at the moment. The late great Helen Keller who was not able to speak, hear, or see became one of the great human inspirations of our time. Not only did she overcome these terrible afflictions and disabilities but became a leading scholar, motivational speaker, role model, author, and thought-provoking leader in her own right. She once said, to quote, **"The only thing worse than being blind is having sight and no vision."** Among other quotes from Ms. Keller are:

> "The best and most beautiful things in the world cannot be seen or even touched. They must be felt with the heart."

"I believe that life is given us so that we may grow in love, and I believe that God is in me as the sun is in the color and fragrance of a flower."

"Keep your face to the sunshine and you cannot see the shadow."

"Faith is the strength by which a shattered world shall emerge into the light."

"I thank God for my handicaps for, through them, I have found myself, my work, and my God."

"When we do the best that we can, we never know what miracle is wrought in our life, or in the life of another."

"God Himself is not secure, having given man dominion over His works."

"To keep our faces toward change, and behave like free spirits in the presence of fate, is strength undefeatable."

"I long to accomplish a great and noble task, but it is my chief duty to accomplish small tasks as if they were great and noble."

"Everything has its wonders, even darkness and silence, and I learn, whatever state I may be in, therein to be content."

"Love is like a beautiful flower which I may not touch, but whose fragrance makes the garden a place of delight just the same."

To conclude, I believe that the Lord has guided me in writing this book and that my words are spontaneous by the power of the hand of He who knows all, the omnipresent and omnipotent Creator whose knowledge and wisdom is the subject of praise by this author. Glory to God forever.

About the Author

Frank Young is a graduate of Georgia State University whose professional career included Georgia Public Television (PBS Network), Georgia State, and Reed Elsevier (global publishers). He has always been awed by Nature and has sought answers to our existence on earth and for what purpose. As a former researcher he is convinced that the best scientific testimony to Creationism are the plethora of Nature's miracles. He is a member of The Georgia Association of Retired Educators and is married to Rosemary Dashiell-Young to whom he has dedicated this book.

CPSIA information can be obtained
at www.ICGtesting.com
Printed in the USA
FSOW01n1656140217
30807FS